SPOTLIGHT ON NATURE

ORANGUTAN

PAMELA DELL

CREATIVE EDUCATION • CREATIVE PAPERBACKS

Published by Creative Education and Creati
P.O. Box 227, Mankato, Minnesota 56002
Creative Education and Creative Paperback
are imprints of The Creative Company
www.thecreativecompany.us

Design and production by Blue Design, Inc.
Art direction by Graham Morgan
Edited by Grace Beltowski

Photographs by Dreamstime/Eko Pramono,
ty Heritage Library, 8, 10, 14, 16, 20, 22; Ge
Cavan Images, 17, Jami Tarris, 18, Jose Azel,
McManus, 9, Thom E. / 500px, 21; Pexels/Florian
Lorenzo Manera, 23; Unsplash/Chris Charles, 11, David
rizio Frigeni, 27, Rob Potter, 12, Simone Dinoia, 15; Wikimed
Arifinal0109, 14, Bernard DUPONT, 29, Calistemon, 28, Eric Kil
Greg Hume, 29, Joeavison1, 10, Malene Thyssen, 16

Library of Congress Cataloging-in-Publication Data
Names: Dell, Pamela, author.
Title: Orangutan / Pamela Dell.
Description: Mankato, Minnesota : Creative Education / Creative Paperbacks, [2026] | Series: Spotlight on nature | Includes bibliographical references and index. | Audience: Ages 10-13 | Audience: Grades 4-6 | Summary: "Climb into the orangutan's world with this nature title for middle-grade wildlife lovers. Informational text pairs with a narrative about a single great ape family to spotlight the wild animal's life cycle, supported by infographics and a timeline of developmental milestones"— Provided by publisher.
Identifiers: LCCN 2024048403 (print) | LCCN 2024048404 (ebook) | ISBN 9798889896265 (library binding) | ISBN 9781682777923 (paperback) | ISBN 9798889897064 (ebook)
Subjects: LCSH: Orangutans—Indonesia—Sumatra—Juvenile literature. | Endangered species—Indonesia—Sumatra—Juvenile literature.
Classification: LCC QL737.P94 D36 2026 (print) | LCC QL737.P94 (ebook) | DDC 599.88/3—dc23/eng/20241220
LC record available at https://lccn.loc.gov/2024048403
LC ebook record available at https://lccn.loc.gov/2024048404

Printed in India

CONTENTS

MEET THE FAMILY

ORANGUTANS of Sumatra

The island of Sumatra is part of Indonesia in Southeast Asia. It is a paradise of deep, humid rainforests. These forests are protected as national parks. Among the rainforest tangle lives a vastly diverse range of plants and animals: 201 **species** of mammals, 580 bird species, and about 10,000 different plant species. The Sumatran tiger is on the prowl here. Elephants and wild boars roam this land. Miniature orchids grow alongside the world's largest flower, the nasty-smelling corpse flower. Hornbills, owls, and pheasants also make their homes here. And high up in the trees lives a rare great ape, the Sumatran orangutan.

Sixty-five feet (19.8 meters) up in a rainforest tree, a Sumatran orangutan has built a strong nest. She has just given birth to a single baby. It is entirely helpless. To survive, it will need her full attention for a long time.

CLOSE-UP

High intelligence

Orangutans are known to be one of the most intelligent **primate** species after humans. One indication of this is in their advanced tool-using and tool-making abilities. Studies show they also have highly developed communication skills. This even includes sign language communication.

CHAPTER ONE

LIFE BEGINS

The enormous hairy creatures known as orangutans are not monkeys. They are great apes. Apes do not have tails and are generally bigger than monkeys. Orangutans are especially massive. In the wild, they are found only on Sumatra and the neighboring island of Borneo. They are also the only great apes in all of Asia.

In the Malay and Indonesian languages, *orang* means "person." *Utan* comes from the word *hutan*, which means "forest." So the orangutan (or *orang-utan* in its home region) is a "person of the forest." These primates are the largest tree-dwelling mammals in the world. Their lives are spent almost entirely in the trees. They move from tree to tree by climbing and swinging.

There are three orangutan species, all with shaggy, gingery-red fur. Two of them are the Bornean and Sumatran orangutans. A second Sumatran species, the Tapanuli orangutan, was newly discovered in 2017.

ORANGUTAN MILESTONES

DAY 1

- Born
- Weak and cannot raise its head
- Weight: 3.3 pounds (1.5 kilograms)

2 WEEKS

- Can sit up and use its hands
- Carried everywhere by mother

Orangutans eat mostly ripe fruit found in the **canopy**. This includes figs, lychees, mangoes, and a favorite, called durian, a spiny, smelly tropical fruit. They also eat young plant shoots and leaves, tree bark, insects, and even soil. They get water from tree holes and the fruits they eat.

Crocodiles, pythons, and clouded leopards are all orangutan predators. But it is usually the young ones they go after. The adults are strong, with long, powerful arms. When standing, their arms reach nearly to the ground. Males weigh between 205 to 287 pounds (93–130 kg). They stand about 5 feet (1.5 m) tall. Females are smaller, around 106 to 121 pounds (48–55 kg), and are about 4 feet (1.2 m) in height.

CLOSE-UP

Tool use

Orangutans are masters of tool use. They sometimes use wet leaves to clean themselves. They have been seen using foliage to make rain hats and leak-proof roofs for their sleeping nests. They also learn to use sticks for various tasks, such as fishing and self-protection.

FEATURED FAMILY

Welcome to the World

It is nighttime in the Sumatran rainforest. The female orangutan lies with her newborn baby in the nest she has built in the upper branches of a balau tree. The infant has been alive for only a few hours. She weighs about 3.3 pounds (1.5 kg) and cannot even lift her head. She has a wrinkled, hairless face, big eyes, and no teeth. Her body also has little hair yet. But her fingers are already strong enough to hold tight to her mother's body as they cuddle together.

An orangutan pregnancy is almost as long as that of a human woman. A female orangutan usually delivers only one **offspring**, called an infant, baby, or child. In rare cases, she will give birth to twins. Motherhood is a fulltime job. A baby orangutan depends entirely on its mother. She will nurse her baby, feed it, and teach it everything it needs to know to make its way in the rainforest world. It will be years before the young orangutan reaches full independence. Until then, the little baby will cling to its mother's body day and night.

3 MONTHS

- Continues to nurse
- Begins eating soft fruit

4 MONTHS

- Teeth begin to form

6 MONTHS

- Shows interest in nest building by adding twigs and leaves to mother's nests

CLOSE-UP
Nesting

Every evening before dark, orangutans build a nightly nest, or sleeping platform, high in the forest canopy. Nests are built by weaving and bending tree branches. Sometimes an ape will swing on the branches first to see if the boughs will hold their weight.

FEATURED FAMILY

First Meal

The tiny Sumatran infant is so small she could fit in the palm of a human hand. But she is alive and well. After her mother cleans her up, it is time for her first meal. She is hungry and cries to be fed. By nature, she latches on to her mother, who nurses her with her nourishing milk. The baby wraps her long, skinny arms around her mother. A close bond begins to form between them. Whenever the infant gets hungry, she will let her mother know by crying for more milk.

Orangutans are the

LARGEST

tree-dwelling mammals in the world. They move from tree to tree by **climbing** and **swinging**.

8 MONTHS

- Spends short times alone but near mother

1 YEAR

- No longer carried everywhere by mother
- Begins venturing farther away alone
- Takes food from mother's hand or begs by putting hand to mother's mouth
- Tries to build its own nests

CLOSE-UP

Flanges

Many sexually mature male orangutans develop long, shaggy fur and large cheek pads called flanges. Once flanges start to grow, it takes about a year for them to fully develop. Studies show that females are more attracted to flanged males than unflanged ones.

CHAPTER TWO

EARLY ADVENTURES

Orangutans are mostly solitary creatures. Males keep to themselves unless they are mating or fighting other males. They do not help raise their offspring. Female orangutans sometimes share territories with other females. They may travel in groups. Females may also spend time with younger orangutans other than their own. But their own young are always with them. Many years pass before orangutan babies are ready to live life by themselves. Humans are the only other mammal to form such long, close ties with their children.

A young orangutan depends on its mother for food and transportation for at least two years. It hangs on tightly to her stomach or side. After it develops more confidence, the baby might ride along on its mother's back. Some orangutans spend five years or more feeding on their mothers' milk. Sometimes they continue to breast feed even up to eight years old. During this time, the mother will not have any other babies. Seven to nine years will pass before she has another.

2 YEARS

- Spends only half its time away from mother
- Stops clinging to mother's chest; begins riding her back

3 YEARS

- No longer carried by mother at all
- Begins to forage for own food

CLOSE-UP

Throat sac

Flanged adult males have a large, flabby "throat sac" below their chins. By inflating the sac, they can make loud roaring sounds known as "long calls." Long calls are used to warn off other males and to attract females for mating.

FEATURED FAMILY

First Fruit

In Sumatra, the mother orangutan is on the hunt for food. It is early morning, still cool enough for this demanding work. She carries her three-month-old daughter against her chest. The baby has just begun to eat soft fruit as well as milk. Her mother plucks a mango and grinds it with her teeth. Then she gives it to her daughter to eat. With every move her mother makes, the baby orangutan learns. After their meal, in the heat of midday, they return to their day nest to nap.

Because the time between the birth of one baby and the next is so long, orangutan populations cannot grow rapidly. But a mother orangutan must be focused on teaching her baby everything about survival. As they travel together, babies closely observe how their mothers build nests and forage through the forest. They learn by example. It is a challenging time. There is much to master before a baby orangutan can live on its own.

(4) YEARS

- Has perfected day-nest building
- Attempts night-nest building
- Starts climbing trees on its own
- Uses tools to get at food

CLOSE-UP

Vocalizations

Orangutans often express their emotions through vocalizing, or making sounds. Besides the long call, orangutans use a "fast call." This is a fast-paced sound used as a threat. Orangutans "kiss-squeak" by pursing their lips together and smacking. Kiss-squeaks signal annoyance.

FEATURED FAMILY

Give It a Try

It is raining in the nighttime forest. At six months old, the Sumatran orangutan clutches her mother's side. Her mother quickly fashions a roof over their sleeping nest. She weaves leaves and branches to make the roof waterproof. Her baby watches closely. They stay dry all night. The next morning, they find a beehive in a nearby tree. The mother shows the young one how to poke a stick into the beehive. The baby tries it. She pokes and twirls the stick until she gets honey. Now she knows two important skills.

BABY ORANGUTANS spend years depending on their mothers for food and transportation.

5 YEARS

- Accomplished at night-nest building

7 YEARS

- Has stopped nursing entirely
- Begins leaving mother for very long periods of time to wander
- Builds all its own nests

CLOSE-UP

Arms and legs

Orangutans' arms are about twice as long as their legs. In adults, they measure up to 7 feet (2.1 m) from one outstretched hand to the other. Their differently built hip joints give their legs much freer movement than humans and other primates.

CHAPTER THREE

LIFE LESSONS

Human children have a lot to learn about life before striking out on their own. They need many skills to become independent. The same is true of orangutans. These big red apes can live up to 50 years in the wild. And they are second only to humans in the length of time they "live at home." What they learn in this time will serve them their whole lives.

Orangutans spend most of their daytime hours foraging for food. But the most important thing a baby orangutan needs to know is what makes a meal. An orangutan's diet is about 60 percent fruit. And any rainforest has thousands of different plant species. A mother shows her infant which are good to eat and which are dangerous. The young must learn not only the location of **edible** plants. Knowing when the plants are ripe enough to eat is also important.

Tool use is an orangutan specialty. Short branches and twigs are good for opening some fruits. They are also useful for digging termites out of tree holes and taking honey from bees' nests. Mothers teach

10 YEARS

- Weight (females): 44–66 pounds (20–30 kg)
- Becomes independent of mother and begins living on its own
- Travels widely and starts establishing its own personal home range

their babies how to build strong nests, too. Every night, orangutans build a new sleeping nest. Occasionally they reuse a nest they have already built, with some added foliage. They also build daytime nests to nap or eat in.

At about 35 years old, some male orangutans begin to grow flanges and a big throat sac beneath their chin. Flanges are wide flaps of fatty tissue on the sides of the face. Unflanged males are usually smaller in size, but they can also reproduce. Sometimes unflanged adults grow flanges and throat sacs later on. No one is sure why. But some studies indicate it may have to do with territory. Once a male has a secure area to call his own, flanges may develop.

Whatever the case, male and female orangutans all need to gain vast

FEATURED FAMILY

This Is How It's Done

The five-year-old orangutan now spends lots of time away from her mother. But she always uses what she has learned. Searching for food, she comes across a durian, her favorite fruit. But durians are hard to break open. Their outsides are covered in spines too prickly for her palms. But her mother showed her how to get to the fruit's juicy insides. She pads the fruit with thick leaves to protect her hands. Then, with all her strength, she tears the durian open. She wins a pain-free reward!

CLOSE-UP

Hands and feet

An orangutan's hands and feet are long and very strong. Their feet look more like another pair of hands. Their big toes look more like thumbs. This allows them to use both hands and feet for eating, grabbing, and climbing through the trees.

(14) YEARS

- Females reach sexual maturity and can reproduce
- Weight (males): 66–110 pounds (30–50 kg)

(18) YEARS

- Males reach sexual maturity

experience as they grow to adulthood. Once independent, males leave and never look back. Females often stake out territory near their mothers. They sometimes even return to visit. But without all they have learned in those early years, no orangutan would survive on its own.

CLOSE-UP

Brachiation

Brachiation (*bray-kee-AY-shuhn*) is the name for the way orangutans move through the treetops. They hook their long fingers around a branch. Then they swing forward, grabbing the next branch with the other hand. This takes them effortlessly from one tree to another.

FEATURED FAMILY

Practice Makes Perfect

The young orangutan has watched her mother build nests since infancy. As she grew, she tried nest building herself. Her early nests were loose and came apart. Her mother builds nests quickly, woven from sturdy, leafy branches. Sometimes she adds a "blanket" or "pillow." The youngster has watched and practiced for seven years now. Today she applies all she has learned. She makes a weave of branches and leaves, tight and strong. She crawls in and it does not give way. Maybe next she will try making a pillow.

Some adult male orangutans grow flanges and a big throat sac beneath their chin.

30 YEARS	35 YEARS	45 YEARS
Females stop reproducing	Some males grow flanges	End of life in the wild

CHAPTER FOUR

SAVE THE RAINFORESTS!

Large numbers of orangutans once lived on islands throughout Southeast Asia. Today, they inhabit the rainforests of only two islands. They are critically endangered. This means there are hardly any of them left. Orangutans are one of the world's species most threatened with **extinction**.

Trying to get a count of these big red apes isn't easy. Experts estimate that only 50,000 to 65,000 are left in the wild. More disturbingly, about 2,000 to 3,000 are killed every year. Those who work with and for orangutans agree: Drastic measures must be taken to change this. Without help, orangutans may be gone within 25 years.

The biggest threat to orangutan survival is **deforestation**. Illegal activities, such as hunting, are another danger. It is also against the law to **poach** orangutans for the pet trade. Most captured babies do not survive. If they do, they grow into untamed adults. Nevertheless, these illegal activities continue.

Orangutans need trees to stay alive. But the Sumatran and the Bornean rainforests have been destroyed at alarming rates. Within the last 20 years, more than 80 percent of orangutan habitat has been wiped out. Much of this deforestation is illegal. But beyond all other causes is the unending destruction of forests for palm oil plantations. Growers in search of high profits clear the standing trees to grow the fruit that produces palm oil. In the United States alone, demand for this oil has tripled in only a few years. Palm oil is found in everything from lipsticks to chocolate nut spreads. Buying products that contain palm oil contributes to loss of orangutan habitat.

Human efforts have recently slowed deforestation in Borneo. Indonesia and Malaysia have begun setting greater restrictions on forest clearing, too. There are also many organizations worldwide that are urgently working to prevent the loss of wild orangutans. Some of this work includes educating Indonesians and others about how to protect orangutans. Some take in injured and orphaned orangutans and help them get ready to return to the wild. The Orangutan Project and Orangutan Foundation International are two very active organizations. Another, the Orangutan Conservancy, also provides pointers on how everyone can help. Saving the Indonesian rainforests means rescuing wild orangutans from certain extinction.

FAMILY ALBUM

SNAPSHOTS

Sumatran orangutans are found in the northwest region of Sumatra. **Tapanuli orangutans** live in the southern part of the island.

Tapanuli are the world's rarest great ape species. There are no more than 800 of them in the wild.

Sumatran orangutans have longer faces and more facial hair than other orangutans.

Tapanuli orangutans look like Sumatran orangutans but with smaller heads. They also have frizzier hair and wider, flatter faces.

Bornean orangutans have darker, shorter fur than the other two species. Their faces are also rounder.

Bornean orangutans' unique digestive systems make them able to digest tough, fibrous plants better than other primates.

Tapanuli orangutans eat some unique things that the others may not. This includes caterpillars and cones from conifer trees.

Male Sumatran orangutans remain with the pregnant female until she has given birth. Other males go back to their solitary life after mating.

Studies have shown that Sumatran orangutans may "grow up" later than Bornean orangutans. Sumatran orangutans take longer to leave their mothers.

Some Bornean orangutans have been seen using handfuls of leaves like napkins to wipe their chins while eating.

Of all orangutans, Bornean males are most likely to come down to the ground. Most females rarely leave the treetops at all.

WORDS to Know

canopy the tree branches and leaves at the highest level of a forest

deforestation intentional clearing of forest land

edible safe to eat

extinction being wiped out of existence

foliage leaves, sticks, branches, and other plant matter

forage search for food

offspring an animal's young

poach to illegally capture or kill animals for profit

primate a mammal with a large brain, hands and feet that can grasp, and eyes facing forward; includes humans, apes, and monkeys

species a group of living things that have shared characteristics and are able to reproduce with one another

LEARN MORE

Books

Gudgeon, Christopher. *Great Apes: Protecting Our Animal Cousins*. Custer, Wash.: Orca Book Publishers, 2025.

Kington, Emily. *Orangutans*. Cornwall, U.K.: Hungry Tomato, 2022.

Spilsbury, Louise. *Saving the Orangutan*. Bridgnorth, U.K.: Cheriton Children's Books, 2023.

Websites

"Orangutan." Britannica Kids.
https://kids.britannica.com/kids/article/orangutan/353572

"Orangutan." National Geographic Kids.
https://kids.nationalgeographic.com/animals/mammals/facts/orangutan

"What Is an Orangutan?" Kids' Canopy by Orangutan Conservancy.
https://kids.orangutan.com

Documentaries

Cordey, Huw, dir. *Secret Lives of Orangutans*. Bristol, U.K.: Silverback Films, 2024.

Musgrave, Rowan, dir. *Orangutan Diary*. London: BBC, 2007.

Porter, Anthony, and Gary Shapiro, hosts. *The Orangutan Podcast*. Santa Monica, Calif.: Orang Utan Republik Foundation, 2022.

Note: Every effort has been made to ensure that any websites listed above were active at the time of publication. However, because of the nature of the Internet, it is impossible to guarantee that these sites will remain active indefinitely or that their contents will not be altered.

Visit

SAN DIEGO ZOO

Say hi to the orangutans, including baby Kaja, at the zoo's Lost Forest exhibit.

2920 Zoo Drive
San Diego, CA 92101

SEPILOK ORANGUTAN REHABILITATION CENTRE

Go to one of the daily feedings for a chance to see rehabilitated orangutans in the wild.

Kabili-Sepilok Forest Reserve
Sandakan, Sabah, Malaysia

SMITHSONIAN'S NATIONAL ZOO

Look straight up to get a glimpse of orangutans moving freely along a high cable between exhibits.

3001 Connecticut Avenue N.W.
Washington, D.C. 20008

ZOO ATLANTA

Witness the orangutans' afternoon fruit feeding and hear male orangutans Benny and Satu show off their long calls.

800 Cherokee Avenue S.E.
Atlanta, GA 30315

INDEX